This book belongs to

_____

For Zoey
My Inspiration

ValueReadsBooks
ISBN: 9798665299495

Copyright 2020 by Cameile Graham
All rights Reserved. No part of this book may be reproduced for use, in any format ,without the permission of the author/publisher.

ISBN: 9798665299495
Printed in the USA
2020

A book to COUNT and WASH your hands

 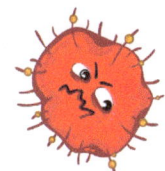 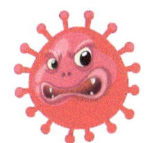

1, 2, 3

NO GERMS ON ME !

Author
CAMEILE GRAHAM

Illustrator
AIWAZ JILANI

Boys and girls, are you staying safe and healthy from germs by washing your hands many times each day?

GREAT !

Here is the FUN way my Dad helps me.

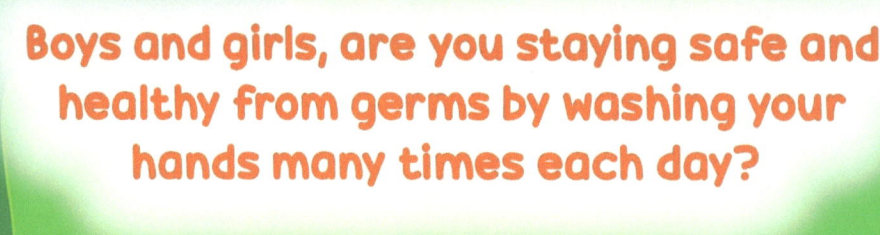

2 TWO

2 drops for two little fingers up to the flow.

6 drops for six little fingers coming closer.

NINE

9 drops for nine little fingers with shine skin.

It is a happy, happy, meal time when I am all done.

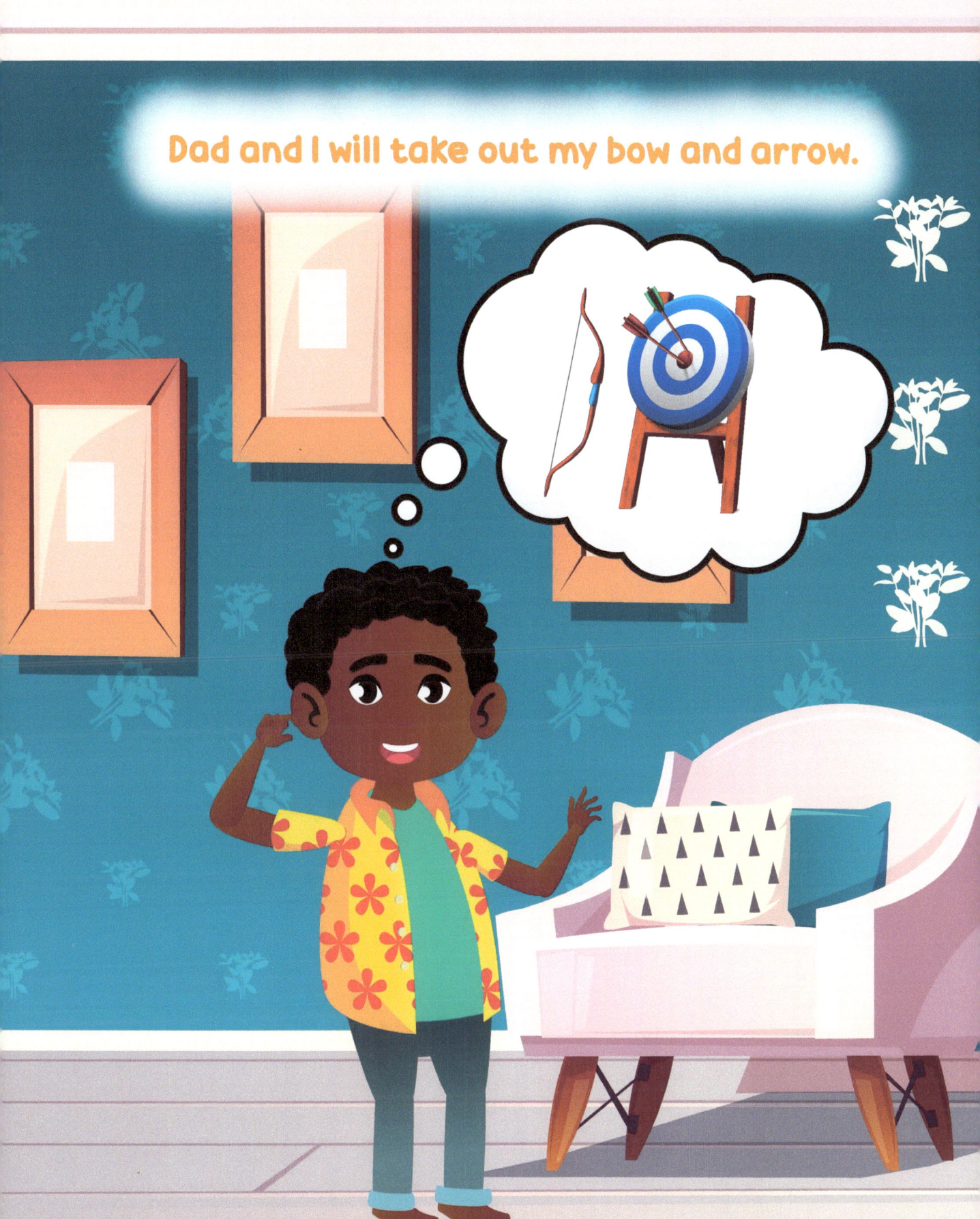

Then counting from 1-10, wash all the germs away, before eating dinner in a safe, clean way.

www.ingramcontent.com/pod-product-compliance
Lightning Source LLC
Chambersburg PA
CBHW051835210526
45473CB00005B/1890